3 Florida FAST Grade 3 Math Practice Tests

Full-Length Test Prep with Detailed Answer Explanations

Dr. A. Nazari

Published by View Math Education

ViewMath.com

3 Practice Tests to Get You Started!

Hey there, future math whiz! ⭐

This book has **3 full practice tests** to help you warm up for the real thing. Think of it like stretching before a big game — these tests will get your brain ready and show you what to expect!

👍 Three tests is the **perfect start**!

👍 Each one helps you feel **more ready**!

👍 You'll be surprised how much you **already know**!

Sharpen your pencil and let's get warmed up! 🔥

> **❝** Three practice tests is a great way to start. Take your time with each one, and you'll feel more confident every step of the way! **❞**

📖 How to Use This Book 📖

Your quick-start guide to 3 great practice tests!

☰ What's Inside This Book

- *3 Full-Length Practice Tests* — Each one covers all the Grade 3 math topics you need to know!

- *Answer Key with Explanations* — Find out why each answer is correct, not just what the answer is.

- *Reference Pages* — A math symbols chart and multiplication table you can peek at any time.

- *A Test Tracker* — Write down your scores and watch your confidence grow!

🗓 A Simple 3-Test Plan

With just 3 tests, here's a great way to use them:

- *Test 1* — **The Warm-Up.** Take this test without a timer. Get comfortable with the question types. Don't worry about your score — just do your best!

- *Test 2* — **The Practice Round.** After reviewing Test 1, try this one with a timer (ask a grown-up!). Focus on the topics that were tricky last time.

- *Test 3* — **The Real Deal.** Treat this like the actual test: quiet room, timed, no peeking at answers. See how much you've improved!

⬤ Multiple Choice

Pick the **one best answer** from choices A, B, C, or D. Not sure? Cross out the ones you know are wrong, then pick from what's left. That's a smart move!

✏ Short Answer

Write your answer **and** show your work! Even if your final answer isn't right, showing your steps can earn you credit. Use scratch paper if you need more room.

66 After Each Test 99

Flip to the Answer Key and check your work. For every question you got wrong, **read the explanation carefully.** Then write the tricky topics on your Test Tracker page. If you need extra help, grab our **Grade 3 Math Study Guide!**

> ⭐ **_Fun fact_**_: Three tests is all it takes to see real improvement! Most kids feel way more confident after just a few rounds of practice._ ⭐

Find more at
ViewMath.com/FL-Grade3

💡 Tips for Test Day 💡

Easy tricks to help you feel calm and do your best!

🌙 The Night Before

- ✅ **Sleep early** — your brain learns while you sleep!
- ✅ **Pack your supplies** — pencils, eraser, scratch paper, all ready to go.
- ✅ **Tell yourself:** "I've been practicing. I'm going to do great!"

👍 5 Simple Rules for Every Test

1. **Read the question twice.** The first time to understand it. The second time to catch details.
2. **Show your work.** Write the steps down, even on scratch paper. It helps you think!
3. **Skip the hard ones.** Put a small star next to tricky questions and come back later. Answer the easy ones first!
4. **Never leave a blank.** For multiple choice, your best guess is better than no answer at all.
5. **Check your work.** Finished early? Go back and re-read your answers.

✅ Smart Moves

- Take a deep breath before you begin
- Underline key words in the question
- Use drawings or number lines to help
- Cross out wrong answers first
- Double-check addition and subtraction

❌ Traps to Avoid

- Rushing and not reading carefully
- Picking the first answer that "looks right"
- Forgetting to carry or borrow numbers
- Skipping a question permanently
- Panicking when you see a tough problem

❝ Remember, the very first practice test is the hardest — not because the questions are harder, but because everything is new! By Test 3, you'll feel like a pro. Trust me! ❞

 Get Ready to Practice 

Pencils

Sharpened and ready!

Eraser

Everyone makes mistakes!

Scratch Paper

For working things out

A Calm Spot

Somewhere quiet to focus

A Grown-Up

To help set a timer

A Can-Do Attitude

You've totally got this!

✔ Allowed During Tests

- Pencils and erasers
- Blank scratch paper
- The **reference pages** in this book
- A ruler (for measurement questions)

✖ Not Allowed

- Calculators
- Phones, tablets, or computers
- Help from anyone else
- Your study guide (save it for after!)

For Parents & Teachers

- With only 3 tests, **space them at least a week apart**. This gives time to review mistakes before trying the next one.
- Let your child take Test 1 untimed to build familiarity.
- After each test, go through the Answer Key together. Focus on **understanding the "why,"** not just the score.
- If a topic keeps tripping them up, review it in our **Grade 3 Math Study Guide** before the next practice test.
- Celebrate every bit of progress — even getting one more question right is a win!

X^1 Math Reference Sheet X^1

You may use this page during your practice tests!

Symbol	Name	What It Means	
$+$	Plus (Add)	Put numbers together.	$3 + 5 = 8$
$-$	Minus (Subtract)	Take away from a number.	$9 - 4 = 5$
$\times$	Times (Multiply)	Add equal groups.	$4 \times 3 = 12$
$\div$	Divide	Split into equal groups.	$12 \div 3 = 4$
$=$	Equals	Both sides are the same.	$2 + 3 = 5$
$>$	Greater Than	The left number is bigger.	$7 > 3$
$<$	Less Than	The left number is smaller.	$2 < 9$
$\frac{1}{2}$	Fraction Bar	Part of a whole.	$\frac{1}{2}$ means 1 out of 2 equal parts

Key Math Words

- **Sum** — the answer when you add
- **Difference** — the answer when you subtract
- **Product** — the answer when you multiply
- **Quotient** — the answer when you divide
- **Factor** — a number you multiply
- **Array** — objects in rows and columns
- **Fraction** — a part of a whole
- **Numerator** — the top number in a fraction
- **Denominator** — the bottom number
- **Equation** — a math sentence with $=$
- **Estimate** — a smart guess, close to the real answer
- **Perimeter** — the distance around a shape
- **Area** — the space inside a shape
- **Rounding** — making a number simpler by going to the nearest ten or hundred

Q Word Problem Clue Words

- **Add** (+): in all, total, altogether, combined, sum, both, more

- **Subtract** (−): how many more, how many left, fewer, difference, remain

- **Multiply** (×): each, every, groups of, times, rows of, per

- **Divide** (÷): share equally, split, each group, how many groups, per

Find more at
ViewMath.com/FL-Grade3

Multiplication Table

×	1	2	3	4	5	6	7	8	9	10	11
1	1	2	3	4	5	6	7	8	9	10	11
2	2	4	6	8	10	12	14	16	18	20	22
3	3	6	9	12	15	18	21	24	27	30	33
4	4	8	12	16	20	24	28	32	36	40	44
5	5	10	15	20	25	30	35	40	45	50	55
6	6	12	18	24	30	36	42	48	54	60	66
7	7	14	21	28	35	42	49	56	63	70	77
8	8	16	24	32	40	48	56	64	72	80	88
9	9	18	27	36	45	54	63	72	81	90	99
10	10	20	30	40	50	60	70	80	90	100	110
11	11	22	33	44	55	66	77	88	99	110	121

How to Use This Table

To find **4 × 7**:

1. Find **4** in the left column (blue).
2. Find **7** in the top row (blue).
3. Follow the row and column until they meet: the answer is **28**!

📈 My Confidence Tracker 📈

Record your scores below. You'll be amazed at your progress!

My name: ________________________

📋 Test	📅 Date	⭐ Score	😊 How I Feel
1		/	
2		/	
3		/	

The easiest topic for me was:

____

The trickiest topic for me was:

____

One thing I got better at from Test 1 to Test 3:

____

Next time I want to try:

____

> *You just finished 3 practice tests — that's awesome! Compare your first score to your last. I bet you'll see real improvement. Ready for more? Check out our 5-test or 7-test books for even more practice!*

Continue Learning at ViewMath Academy!

For Parents, Teachers & Students

Great job on the practice tests! Want to keep improving? ViewMath Academy is your **free online companion** to this book.

- **Score Analyzer** — Enter your answers and instantly see which topics need more practice
- **Interactive Lessons** — Review the concepts behind each question with clear explanations
- **Adaptive Quizzes** — Practice your weak topics with questions that match your level
- **Progress Tracking** — See your mastery grow across all Grade 3 math topics
- **Personalized Dashboard** — A learning plan tailored just for you

Scan to visit ViewMath Academy

viewmath.com/academy

 Free to use · No downloads required · Works on any device 

 Table of Contents

Here's what we'll explore together!

 *Let's learn and have fun!*

Practice Test 1

 30 Questions

✏ Before You Start ✏

- ✔ **Read each question carefully** before choosing your answer.
- ✔ **Show your work** on scratch paper when you need to.
- ✔ **Skip hard questions** and come back to them later.
- ✔ **Check your answers** when you're done.
- ✔ **Take your time** — there's no rush!

★ You've Got This! ★

Do your best and show what you know!

1. *Write 5,600 in expanded form.*

Your Answer

2. *Which number rounds to 500 when rounded to the nearest 100?*

(A) 428

(B) 449

(C) 462

(D) 551

3. *Each place value is how many times bigger than the place to its right?*

(A) 2 times

(B) 5 times

(C) 10 times

(D) 100 times

4. *A city has 85,420 people. What is the value of the digit 5 in this number?*

(A) 5

(B) 500

(C) 5,000

(D) 50,000

5. *What is $506 + 398$ using the standard algorithm?*

(A) 894

(B) 804

(C) 904

(D) 814

6. *A farmer had 635 apples. He sold 278. How many apples does he have left?*

Your Answer

Find more at
ViewMath.com/FL-Grade3

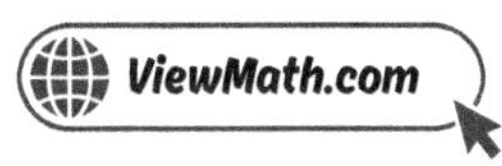

7. Find the sum: $8,456 + 1,544$. Does the answer have 4 or 5 digits? Explain.

Your Answer:

8. When you estimate $347 + 582$ by rounding each number to the nearest hundred, the estimate is 900. The exact sum is 929. Which statement is true?

(A) The estimate is greater than the exact sum

(B) The estimate is less than the exact sum

(C) The estimate equals the exact sum

(D) You cannot compare an estimate to an exact answer

9. What is 8×50?

Your Answer:

10. Which equation shows the identity property of multiplication?

(A) $7 \times 0 = 0$

(B) $7 \times 1 = 7$

(C) $7 \times 2 = 14$

(D) $3 \times 7 = 7 \times 3$

11. Sara has 36 stickers. She shares them equally among 9 friends. How many stickers does each friend get?

(A) 3

(B) 4

(C) 5

(D) 27

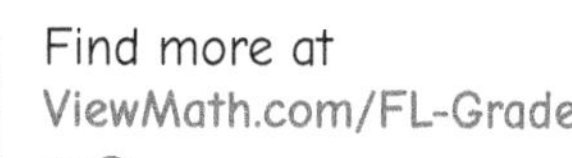

Find more at
ViewMath.com/FL-Grade3

12. Two classes collect cans. Class A collects 35 cans. Class B collects 21 cans. They put the cans in groups of 7 for recycling. How many groups are there?

(A) 5

(B) 7

(C) 8

(D) 56

13. What is 12×4?

14. Which fraction means "two-thirds"?

(A) $\frac{3}{2}$

(B) $\frac{2}{3}$

(C) $\frac{1}{3}$

(D) $\frac{2}{2}$

15. A number line is divided into 8 equal parts. How many marks from 0 is $\frac{7}{8}$?

Your Answer:

16. $\frac{7}{8}$ is built from how many copies of $\frac{1}{8}$?

(A) 1

(B) 7

(C) 8

(D) 15

17. Which fraction is NOT equivalent to $\frac{2}{3}$?

(A) $\frac{4}{6}$

(B) $\frac{6}{8}$

(C) $\frac{6}{9}$

(D) $\frac{8}{12}$

Find more at
ViewMath.com/FL-Grade3

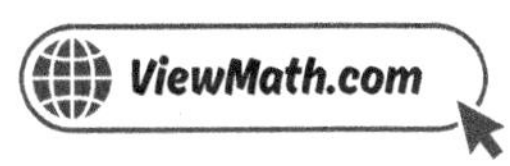

18. Is $\frac{2}{6}$ more or less than $\frac{1}{2}$?

Your Answer:

19. The minute hand points to the 12 and the hour hand points to the 5. What time is it?

(A) 12:05

(B) 12:25

(C) 5:12

(D) 5:00

20. 9,000 grams = how many kilograms?

(A) 9 kg

(B) 90 kg

(C) 900 kg

(D) 9,000 kg

21. A pitcher holds 8 liters of juice. The family drinks 3 liters. How much juice is left?

(A) 3 L

(B) 5 L

(C) 8 L

(D) 11 L

22. You have 3 quarters, 2 dimes, and 1 nickel. How much money do you have in cents?

Your Answer:

23. What is $\$2.50 + \$1.25$?

(A) \$3.25

(B) \$3.75

(C) \$4.75

(D) \$3.55

24. A picture graph has the key: Each ★ = 5 books. The row for Jake has 3 stars. How many books did Jake read?

(A) 3

(B) 5

(C) 8

(D) 15

25. What type of angle is smaller than a right angle?

(A) Obtuse angle

(B) Right angle

(C) Acute angle

(D) Straight angle

26. How many vertices does a cube have?

(A) 4

(B) 6

(C) 8

(D) 12

27. A rectangle has 1 row and 9 columns of unit squares. What is the area?

Your Answer

28. A circle is partitioned into 8 equal parts. If 3 parts are shaded, what fraction of the circle is shaded?

(A) $\frac{3}{3}$

(B) $\frac{3}{8}$

(C) $\frac{8}{3}$

(D) $\frac{1}{8}$

29. An angle is formed by two rays that share the same:

(A) Line

(B) Side

(C) Vertex

(D) Segment

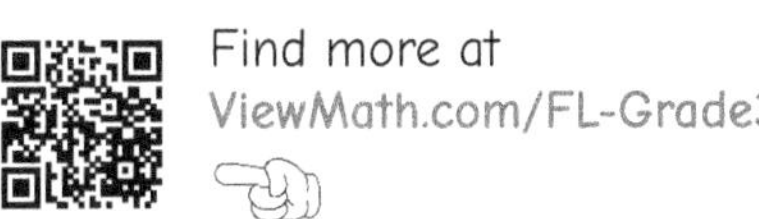
Find more at
ViewMath.com/FL-Grade3

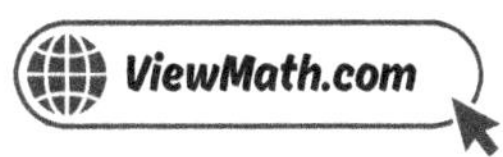

30. Which of these is a real-world example of symmetry?

(A) A crumpled piece of paper

(B) A butterfly

(C) A cloud

(D) A tree branch

Find more at
ViewMath.com/FL-Grade3

 *End of Practice Test 1* 

Great job finishing the test!

 **My Score**

I got _________ out of 30 questions right.

Check your answers in the Answer Key at the back of the book.

 *Review any questions you missed. That's how we learn!*

📊 Check Your Score Online!

Visit **ViewMath Academy** to enter your answers and see which topics you need to review. You can also explore lessons, take quizzes, track your scores, and save your progress!

viewmath.com/score/3.1.FL.01

Or go to viewmath.com/score and enter code: 3.1.FL.01

2

Practice Test 2

30 Questions

✏️ Before You Start ✏️

- ✔ **Read each question carefully** before choosing your answer.
- ✔ **Show your work** on scratch paper when you need to.
- ✔ **Skip hard questions** and come back to them later.
- ✔ **Check your answers** when you're done.
- ✔ **Take your time** — there's no rush!

 **You've Got This!** 

Do your best and show what you know!

1. Write the expanded form of 9,205.

   Your Answer:

2. Maria rounded 438 to the nearest 10 and got 430. Is she correct?

   (A) No, it should be 440

   (B) Yes, 438 rounds to 430

   (C) No, it should be 400

   (D) No, it should be 450

3. What is the largest 4-digit number?

   Your Answer:

4. Write the number $60,000 + 3,000 + 200 + 10 + 5$ in standard form.

   Your Answer:

5. Use the standard algorithm to find $463 + 379$.

   Your Answer:

6. What is $413 - 268$?

   (A) 155

   (B) 245

   (C) 145

   (D) 235

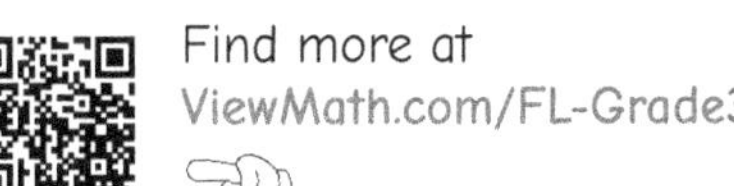

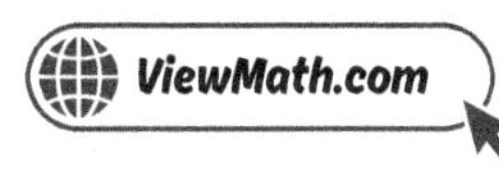

7. Can two 4-digit numbers add up to a 5-digit number?

 (A) No, never.

 (B) Yes, when the thousands column has a carry.

 (C) Only with regrouping in every column.

 (D) Only if both numbers are 9,999.

8. Estimate $3,412 + 2,678$ by rounding each number to the nearest thousand.

 (A) 5,000

 (B) 6,000

 (C) 7,000

 (D) 6,090

9. What is 6×70?

 (A) 42

 (B) 130

 (C) 420

 (D) 4,200

10. Which statement is TRUE?

 (A) $0 \times 5 = 5$

 (B) $1 \times 5 = 0$

 (C) $0 \times 5 = 0$

 (D) $1 \times 0 = 1$

11. A florist makes bouquets with 6 flowers each. She has 54 flowers. How many bouquets can she make?

 (A) 7

 (B) 8

 (C) 9

 (D) 48

12. Emma buys 3 packs of stickers with 8 stickers each. She gives 5 stickers to her friend. How many stickers does Emma have left?

 (A) 16

 (B) 19

 (C) 21

 (D) 24

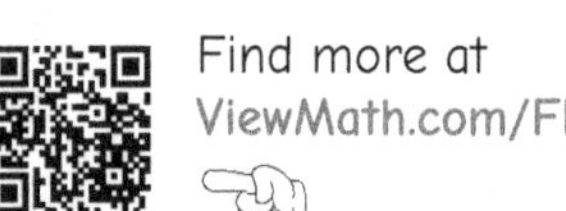

13. *What is $77 \div 11$?*

   (A) 6   (B) 7

   (C) 8   (D) 11

14. *Lucy says $\frac{5}{4}$ means 5 parts out of 4. Is that possible?*

   (A) *Yes, you just need more than one whole*   (B) *No, the numerator can never be bigger than the denominator*

   (C) *No, fractions always equal less than 1*   (D) *Yes, but only with circles*

15. *A number line is divided into 4 equal parts. Which fraction is at the halfway point between 0 and 1?*

   (A) $\frac{1}{4}$ (B) $\frac{2}{4}$

   (C) $\frac{3}{4}$ (D) $\frac{4}{4}$

16. *How many copies of $\frac{1}{3}$ make 1 whole?*

   (A) 1   (B) 2

   (C) 3   (D) 4

17. *$\frac{3}{4} = \frac{?}{8}$. What is the missing numerator?*

   (A) 3   (B) 4

   (C) 6   (D) 7

18. *Is $\frac{5}{8}$ more or less than $\frac{1}{2}$? Explain.*

Your Answer:

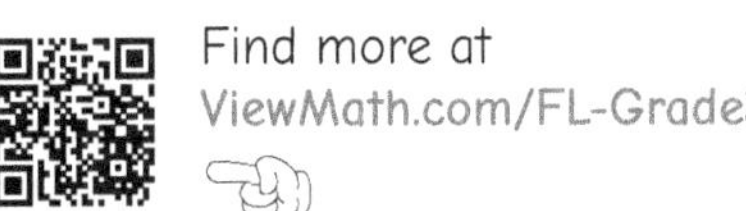

19. Write the time for "quarter past 11" using numbers.

Your Answer:

20. A bag of rice has a mass of 5 kg. You use 2 kg for cooking. How much rice is left?

Your Answer:

21. Which container holds the LEAST liquid?

(A) A swimming pool

(B) A bathtub

(C) A bucket

(D) A teaspoon

22. Which group of coins equals 50 cents?

(A) 5 pennies

(B) 5 dimes

(C) 3 dimes

(D) 3 quarters

23. Lily has $6.00. She buys a sticker pack for $2.35 and a pen for $1.80. Does she have enough money left to buy a snack for $2.00?

(A) Yes, she has $2.85 left

(B) Yes, she has $2.00 left

(C) No, she only has $1.85 left

(D) No, she only has $1.65 left

24. A picture graph tracks books read in a month. Each symbol stands for 2 books. Amy has 5 symbols, Ben has 8 symbols, and Chloe has 3 symbols. How many books did they read altogether?

Your Answer:

Find more at
ViewMath.com/FL-Grade3

25. Which statement about a rectangle is TRUE?

(A) All 4 sides are always equal.   (B) It has exactly 3 right angles.

(C) It has 4 right angles and 2 pairs of equal sides.   (D) It has no parallel sides.

26. Which 3D shape has 1 flat circular face and comes to a point at the top?

(A) Cylinder   (B) Sphere

(C) Cube   (D) Cone

27. A rectangle has 4 rows and 6 columns of unit squares. What is the area?

Your Answer:

28. A shape is partitioned so that each part is $\frac{1}{8}$ of the whole. How many equal parts is the shape split into?

Your Answer:

29. The hands of a clock at 3 o'clock form what type of angle?

(A) Acute angle   (B) Right angle

(C) Obtuse angle   (D) Straight angle

30. Name one real-world object that has symmetry.

Your Answer:

Find more at
ViewMath.com/FL-Grade3

 # End of Practice Test 2 

Great job finishing the test!

☑ My Score

I got _____________ out of 30 questions right.

*Check your answers in the **Answer Key** at the back of the book.*

💡 Review any questions you missed. That's how we learn!

📊 Check Your Score Online!

Visit **ViewMath Academy** to enter your answers and see which topics you need to review. You can also explore lessons, take quizzes, track your scores, and save your progress!

viewmath.com/score/3.1.FL.02

Or go to viewmath.com/score and enter code: 3.1.FL.02

3

Practice Test 3

☑ 30 Questions

✏️ Before You Start ✏️

- ✔ **Read each question carefully** before choosing your answer.
- ✔ **Show your work** on scratch paper when you need to.
- ✔ **Skip hard questions** and come back to them later.
- ✔ **Check your answers** when you're done.
- ✔ **Take your time** — there's no rush!

⭐ You've Got This! ⭐

Do your best and show what you know!

1. *Each place value is how many times bigger than the place to its right?*

(A) 2 times

(B) 5 times

(C) 10 times

(D) 100 times

2. *When you round 549 to the nearest 10 and to the nearest 100, which gives the larger answer?*

(A) Rounded to the nearest 10

(B) Rounded to the nearest 100

(C) They are the same

(D) Cannot tell without rounding

3. *What number is 1 more than 9,999?*

(A) 9,000

(B) 9,990

(C) 10,000

(D) 10,001

4. *Which is the expanded form of 72,385?*

(A) $7,000 + 2,000 + 300 + 80 + 5$

(B) $70,000 + 2,000 + 300 + 80 + 5$

(C) $72,000 + 385$

(D) $70,000 + 2,000 + 380 + 5$

5. *What is $648 + 352$?*

(A) 900

(B) 990

(C) 1,000

(D) 1,010

6. *What is $524 - 367$ using the standard algorithm?*

(A) 157

(B) 167

(C) 257

(D) 147

  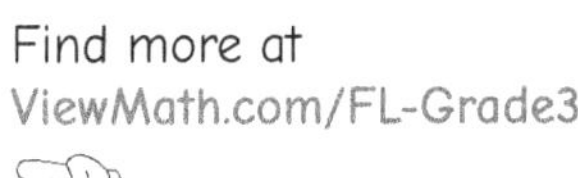
Find more at
ViewMath.com/FL-Grade3

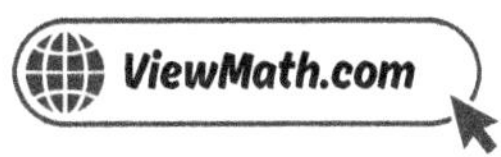

7. Maya added $4{,}839 + 2{,}764$ and got $7{,}503$. She estimates $4{,}800 + 2{,}800 = 7{,}600$. What should Maya do?

(A) Accept her answer since it is close to the estimate.

(B) Redo the problem because the estimate is too far away.

(C) Round both numbers to the nearest thousand instead.

(D) Subtract to check.

8. Estimate $724 - 389$ by rounding each number to the nearest hundred.

(A) 200

(B) 300

(C) 400

(D) 335

9. A movie theater has 8 rows of seats with 20 seats in each row. How many seats are there?

(A) 28

(B) 160

(C) 180

(D) 1,600

10. A student says: "$0 \times 9 = 9$ because zero doesn't change anything." What mistake did the student make?

(A) The student is correct

(B) The student confused the zero property with the identity property; $0 \times 9 = 0$

(C) The student should have added instead of multiplied

(D) The answer should be 90

11. Which equation solves this problem? "There are 35 stickers shared equally among 5 friends."

(A) $35 + 5 = 40$

(B) $35 - 5 = 30$

(C) $35 \times 5 = 175$

(D) $35 \div 5 = 7$

12. There are 27 boys and 27 girls on a field trip. They split into 6 equal groups. How many students are in each group?

Your Answer

13. What is 11×12?

(A) 112

(B) 121

(C) 132

(D) 144

14. Which fraction has a numerator of 2 and a denominator of 6?

(A) $\frac{6}{2}$

(B) $\frac{2}{6}$

(C) $\frac{2}{2}$

(D) $\frac{6}{6}$

15. Is $\frac{3}{3}$ at 0 or at 1 on the number line?

Your Answer

16. On a number line, you are at $\frac{2}{8}$. You take 3 more hops of $\frac{1}{8}$. Where do you land?

(A) $\frac{3}{8}$

(B) $\frac{5}{8}$

(C) $\frac{5}{16}$

(D) $\frac{6}{8}$

17. Find a fraction equivalent to $\frac{1}{3}$ by multiplying the numerator and denominator by 2.

(A) $\frac{2}{3}$

(B) $\frac{1}{6}$

(C) $\frac{2}{6}$

(D) $\frac{3}{6}$

Find more at
ViewMath.com/FL-Grade3

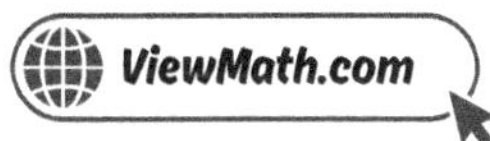

18. Ruby says $\frac{1}{8} > \frac{1}{3}$ because $8 > 3$. Is Ruby correct?

(A) Yes, bigger numbers are always greater

(B) No, $\frac{1}{3} > \frac{1}{8}$ because thirds are bigger pieces than eighths

(C) They are equal

(D) Yes, eighths are bigger than thirds

19. The short hand is past the 7 and the long hand is on the 10, then 2 more tick marks past the 10. What time is it?

(A) 7:50

(B) 7:52

(C) 10:37

(D) 7:10

20. A box of cereal has a mass of 500 g. You buy 2 boxes. What is the total mass?

(A) 502 g

(B) 700 g

(C) 1,000 g

(D) 5,000 g

21. A pot holds 6 liters when full. You pour in 2 liters first, then 3 more liters. How many more liters are needed to fill the pot?

Your Answer:

22. How much is a quarter worth?

(A) 1 cent

(B) 5 cents

(C) 10 cents

(D) 25 cents

23. What is $3.45 + $2.80?

(A) $5.25

(B) $6.25

(C) $5.65

(D) $6.15

24. A picture graph shows toys collected. Each symbol stands for 5 toys.

Bears: ★★★
Cars: ★★★★★
Dolls: ★★

How many toys were collected in all?

(A) 10

(B) 25

(C) 50

(D) 75

25. A shape has 8 sides and 8 vertices. What is the name of this shape?

Your Answer:

26. Lily says a cube and a rectangular prism are exactly the same shape. Is she correct?

(A) Yes, they both have 6 faces.

(B) Yes, they both have 12 edges.

(C) No, a cube has all square faces but a rectangular prism can have different-sized rectangle faces.

(D) No, a rectangular prism has more vertices.

27. Which two rectangles have the same area?

(A) 2×5 and 3×4

(B) 3×6 and 2×9

(C) 4×5 and 3×7

(D) 1×8 and 2×5

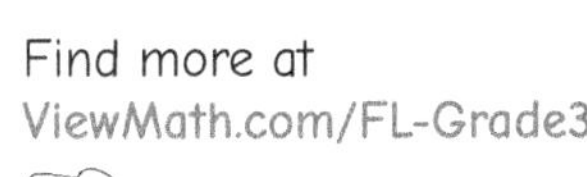
Find more at
ViewMath.com/FL-Grade3

28. If one pizza is cut into 4 equal slices and another identical pizza is cut into 8 equal slices, which slices are bigger?

(A) The $\frac{1}{8}$ slices

(B) The $\frac{1}{4}$ slices

(C) They are the same size.

(D) You cannot tell.

29. How many right angles can you see in the capital letter "L"?

(A) 0

(B) 1

(C) 2

(D) 4

30. How many lines of symmetry does an isosceles triangle have?

(A) 0

(B) 1

(C) 2

(D) 3

Find more at
ViewMath.com/FL-Grade3

 # End of Practice Test 3 

Great job finishing the test!

✅ My Score

I got _____________ out of 30 questions right.

Check your answers in the **Answer Key** at the back of the book

💡 Review any questions you missed. That's how we learn!

📊 Check Your Score Online!

Visit **ViewMath Academy** to enter your answers and see which topics you need to review. You can also explore lessons, take quizzes, track your scores, and save your progress!

viewmath.com/score/3.1.FL.03

Or go to **viewmath.com/score** and enter code: 3.1.FL.03

Answer Key & Explanations

 **Check Your Answers!** 

First try each test on your own, then look here to check.

Read the explanations to learn from any mistakes

Practice Test 1 — Answer Key

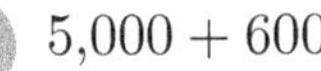 5,000 + 600    C    C    C    C    357

 10,000. *It has 5 digits because the thousands column produces a carry.*    B    400   10 B

11 B   12 C   13 48   14 B   15 7   16 B   17 B   18 Less   19 D   20 A

21 B   22 100 *cents (or $1.00)*   23 B   24 D   25 C   26 C   27 9 *square units*

28 B   29 C   30 B

 **Time to Learn!** 

*Go through the explanations below, **especially for the questions you missed**.*

Understanding why each answer is correct makes you a stronger math thinker!

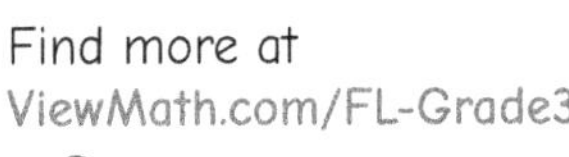 **Tip:** *Circle any questions you got wrong, then read their explanation carefully.*

Practice Test 1 — Detailed Explanations

Find more at
ViewMath.com/FL-Grade3

 ViewMath.com

1. $5,600 = 5,000 + 600 + 0 + 0$. *The tens and ones are both 0.*

2. *462 has tens digit* $6 \geq 5$, *so it rounds up to 500.* 428 *and* 449 *round to* 400. 551 *rounds to* 600.

3. *Each place value is 10 times bigger than the one to its right: ones $\rightarrow$ tens $\rightarrow$ hundreds $\rightarrow$ thousands $\rightarrow$ ten-thousands.*

4. *In 85,420, the digit 5 is in the thousands place, so its value is 5,000.*

5. *Ones:* $6 + 8 = 14$. *Write 4, carry 1. Tens:* $0 + 9 + 1 = 10$. *Write 0, carry 1. Hundreds:* $5 + 3 + 1 = 9$. *The sum is* 904.

6. *Ones:* $5 < 8$. *Borrow:* $15 - 8 = 7$. *Tens:* $2 < 7$. *Borrow:* $12 - 7 = 5$. *Hundreds:* $5 - 2 = 3$. *The farmer has 357 apples left.*

7. *Ones:* $6 + 4 = 10$, *write 0, carry 1. Tens:* $5 + 4 + 1 = 10$, *write 0, carry 1. Hundreds:* $4 + 5 + 1 = 10$, *write 0, carry 1. Thousands:* $8 + 1 + 1 = 10$. *Sum:* 10,000, *a 5-digit number.*

8. $347 \approx 300$ *and* $582 \approx 600$. *The estimate is* $300 + 600 = 900$, *which is less than the exact sum of* 929.

9. *Basic fact:* $8 \times 5 = 40$. *Add one zero:* 400.

10. *The identity property says multiplying any number by 1 gives that same number.* $7 \times 1 = 7$.

11. *Sharing equally means division.* $36 \div 9 = 4$ *stickers each.*

12. *Step 1:* $35 + 21 = 56$ *cans. Step 2:* $56 \div 7 = 8$ *groups.*

13. $(10 \times 4) + (2 \times 4) = 40 + 8 = 48$.

Find more at
ViewMath.com/FL-Grade3

14 *"Two-thirds" means 2 parts out of 3 equal parts* $= \frac{2}{3}$.

15 $\frac{7}{8}$ *is 7 marks from 0 because the numerator counts the marks.*

16 $\frac{7}{8} = 7$ *copies of* $\frac{1}{8}$.

17 $\frac{4}{6} = \frac{2}{3}$, $\frac{6}{9} = \frac{2}{3}$, *and* $\frac{8}{12} = \frac{2}{3}$. *But* $\frac{6}{8} = \frac{3}{4}$, *which is not equivalent to* $\frac{2}{3}$.

18 $\frac{1}{2} = \frac{3}{6}$. *Since* $\frac{2}{6} < \frac{3}{6}$, *it is less than* $\frac{1}{2}$.

19 *When the minute hand points to 12, it means 0 minutes (o'clock). The hour hand on 5 means the hour is 5. The time is 5:00.*

20 *Divide by 1,000:* $9,000 \div 1,000 = 9$ *kg.*

21 $8 - 3 = 5$ *L.*

22 $3 \times 25 = 75$. *Then* $2 \times 10 = 20$. *Then* $1 \times 5 = 5$. *Total:* $75 + 20 + 5 = 100$ *cents.*

23 *Line up the decimals and add:* $\$2.50 + \$1.25 = \$3.75$.

24 *Each star = 5 books. Jake has 3 stars, so* $3 \times 5 = 15$ *books.*

25 *An acute angle is smaller than a right angle. An obtuse angle is larger than a right angle.*

26 *A cube has 8 vertices (corners). Vertices are the points where edges meet.*

27 $1 \times 9 = 9$ *square units.*

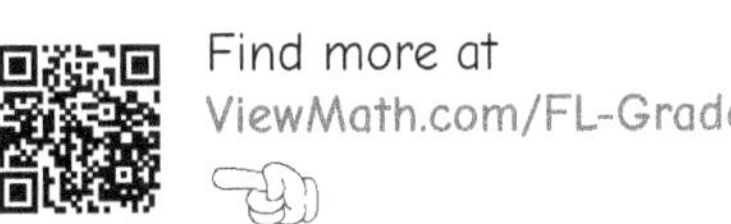

Find more at
ViewMath.com/FL-Grade3

28. The circle has 8 equal parts and 3 are shaded. The fraction shaded is $\frac{3}{8}$.

29. An angle is formed by two rays sharing the same endpoint, called the vertex.

30. A butterfly has a line of symmetry down the middle — the left wing matches the right wing.

✅ Practice Test 2 — Answer Key

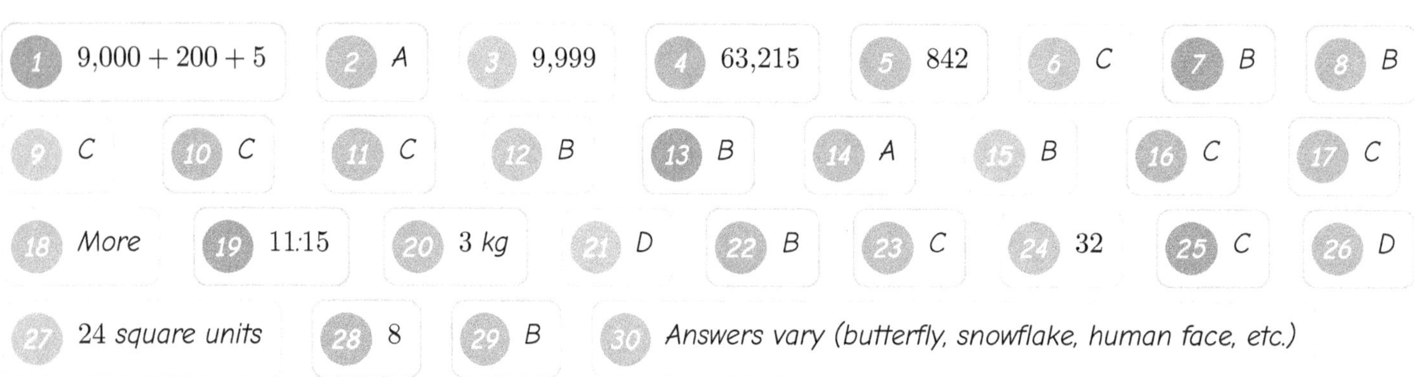

1. $9{,}000 + 200 + 5$
2. A
3. 9,999
4. 63,215
5. 842
6. C
7. B
8. B
9. C
10. C
11. C
12. B
13. B
14. A
15. B
16. C
17. C
18. More
19. 11:15
20. 3 kg
21. D
22. B
23. C
24. 32
25. C
26. D
27. 24 square units
28. 8
29. B
30. Answers vary (butterfly, snowflake, human face, etc.)

💡 Time to Learn! 💡

Go through the explanations below, **especially for the questions you missed.**

Understanding why each answer is correct makes you a stronger math thinker!

👍 **Tip:** Circle any questions you got wrong, then read their explanation carefully.

📖 Practice Test 2 — Detailed Explanations

1. $9{,}205 = 9{,}000 + 200 + 0 + 5$. The tens digit is 0, so there is no tens term.

2. The ones digit is 8. Since $8 \geq 5$, we round up. 438 rounds to 440, not 430.

3. The largest 4-digit number is 9,999. The next number, 10,000, has 5 digits.

4. $60,000 + 3,000 + 200 + 10 + 5 = 63,215$.

5. Ones: $3 + 9 = 12$. Write 2, carry 1. Tens: $6 + 7 + 1 = 14$. Write 4, carry 1. Hundreds: $4 + 3 + 1 = 8$. The sum is 842.

6. Ones: $3 < 8$. Borrow: $13 - 8 = 5$. Tens: $0 < 6$. Borrow: $10 - 6 = 4$. Hundreds: $3 - 2 = 1$. The difference is 145.

7. Yes! If the thousands column (plus any carry) totals 10 or more, the sum is a 5-digit number. For example, $6,500 + 4,500 = 11,000$.

8. $3,412 \approx 3,000$ and $2,678 \approx 3,000$. So $3,000 + 3,000 = 6,000$.

9. Basic fact: $6 \times 7 = 42$. Add one zero back: 420.

10. The zero property says $0 \times 5 = 0$. Choice A confuses the zero and identity properties.

11. $54 \div 6 = 9$ bouquets.

12. Step 1: $3 \times 8 = 24$ stickers. Step 2: $24 - 5 = 19$ stickers left.

13. $11 \times 7 = 77$, so $77 \div 11 = 7$.

14. $\frac{5}{4}$ means 5 parts where each part is $\frac{1}{4}$. That is more than one whole $(1\frac{1}{4})$.

15 $\frac{2}{4}$ is at the second of 4 marks, which is the middle of the line from 0 to 1.

16 $\frac{1}{3} + \frac{1}{3} + \frac{1}{3} = \frac{3}{3} = 1$ whole. It takes 3 copies.

17 The denominator went from 4 to 8 (multiplied by 2). So multiply the numerator by 2 too: $3 \times 2 = 6$.

18 $\frac{1}{2} = \frac{4}{8}$. Since $\frac{5}{8} > \frac{4}{8}$, it is more than $\frac{1}{2}$.

19 "Quarter past" means 15 minutes after the hour. Quarter past 11 is 11:15.

20 $5 - 2 = 3$ kg.

21 A teaspoon holds much less than 1 liter, which is far less than a bucket, bathtub, or pool.

22 5 dimes $= 5 \times 10 = 50$ cents.

23 Total spent: $\$2.35 + \$1.80 = \$4.15$. Money left: $\$6.00 - \$4.15 = \$1.85$. Since $\$1.85 < \$2.00$, she does not have enough.

24 Amy: $5 \times 2 = 10$. Ben: $8 \times 2 = 16$. Chloe: $3 \times 2 = 6$. Total: $10 + 16 + 6 = 32$ books.

25 A rectangle has 4 right angles and 2 pairs of equal sides (opposite sides are equal). Not all 4 sides have to be equal.

26 A cone has 1 flat circular face on the bottom and 1 curved surface that comes to a point (vertex) at the top.

27 4 rows $\times$ 6 columns $= 24$ square units.

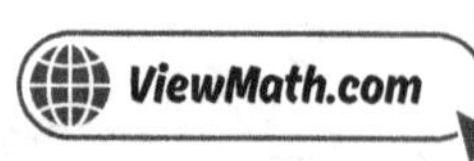

(28) The denominator of $\frac{1}{8}$ is 8, which tells you the shape is split into 8 equal parts.

(29) At 3 o'clock the minute hand points to 12 and the hour hand points to 3, forming a 90° right angle.

(30) Many real-world objects have symmetry, such as butterflies, snowflakes, and starfish.

✔ Practice Test 3 — Answer Key

1 C	**2** A	**3** C	**4** B	**5** C	**6** A	**7** D	**8** B	**9** B	**10** B
11 D	**12** 9	**13** C	**14** B	**15** 1	**16** B	**17** C	**18** B	**19** B	**20** C
21 1 L	**22** D	**23** B	**24** C	**25** Octagon	**26** C	**27** B	**28** B	**29** B	
30 B									

💡 Time to Learn! 💡

Go through the explanations below, **especially for the questions you missed**.

Understanding why each answer is correct makes you a stronger math thinker!

👍 **Tip:** Circle any questions you got wrong, then read their explanation carefully.

📖 Practice Test 3 — Detailed Explanations

(1) Each place is 10 times bigger than the place to its right: ones → tens → hundreds → thousands.

2 549 rounded to the nearest 10: ones digit is $9 \geq 5$, so 550. Rounded to the nearest 100: tens digit is $4 < 5$, so 500. $550 > 500$.

3 $9{,}999 + 1 = 10{,}000$. This is the first 5-digit number.

4 $72{,}385 = 70{,}000 + 2{,}000 + 300 + 80 + 5$. The 7 is in the ten-thousands place worth 70,000.

5 Ones: $8 + 2 = 10$. Write 0, carry 1. Tens: $4 + 5 + 1 = 10$. Write 0, carry 1. Hundreds: $6 + 3 + 1 = 10$. Write 0, carry 1 to thousands. The sum is 1,000.

6 Ones: $4 < 7$. Borrow: $14 - 7 = 7$. Tens: $1 < 6$. Borrow: $11 - 6 = 5$. Hundreds: $4 - 3 = 1$. The difference is 157.

7 The estimate 7,600 is close to 7,503. To be sure, Maya should check by subtracting: $7{,}503 - 2{,}764$ should equal 4,839. The correct sum is actually 7,603, so she has an error.

8 $724 \approx 700$ and $389 \approx 400$. So $700 - 400 = 300$.

9 $8 \times 20 = 160$. Basic fact: $8 \times 2 = 16$, add one zero: 160.

10 The zero property says $0 \times 9 = 0$. The student confused it with the identity property, where $1 \times 9 = 9$.

11 Sharing equally means dividing. $35 \div 5 = 7$ stickers per friend.

12 Step 1: $27 + 27 = 54$ students. Step 2: $54 \div 6 = 9$ students per group.

13 $11 \times 12 = (11 \times 10) + (11 \times 2) = 110 + 22 = 132$.

14 Numerator is on top and denominator is on the bottom. So it is $\frac{2}{6}$.

Find more at
ViewMath.com/FL-Grade3

15 $\frac{3}{3}$ means all 3 parts, which equals 1 whole.

16 $\frac{2}{8} + 3$ hops of $\frac{1}{8} = \frac{2}{8} + \frac{3}{8} = \frac{5}{8}$.

17 $\frac{1 \times 2}{3 \times 2} = \frac{2}{6}$. So $\frac{1}{3} = \frac{2}{6}$.

18 With the same numerator, the bigger denominator gives **smaller** pieces. $\frac{1}{3} > \frac{1}{8}$.

19 The hour is 7. The long hand on 10 means 50 minutes. Two more tick marks gives $50 + 2 = 52$ minutes. The time is 7:52.

20 $500 + 500 = 1,000$ g (which is the same as 1 kg).

21 You poured in $2 + 3 = 5$ L. The pot holds 6 L, so $6 - 5 = 1$ more liter is needed.

22 A quarter is worth 25 cents.

23 Cents: $45 + 80 = 125$ cents $= 1$ dollar and 25 cents. Dollars: $3 + 2 + 1 = 6$. Answer: $6.25.

24 Bears: $3 \times 5 = 15$. Cars: $5 \times 5 = 25$. Dolls: $2 \times 5 = 10$. Total: $15 + 25 + 10 = 50$ toys.

25 A polygon with 8 sides is called an octagon. "Oct" means 8.

26 Both have 6 faces, 12 edges, and 8 vertices. But a cube's faces are all equal squares, while a rectangular prism's faces can be different-sized rectangles.

27 $3 \times 6 = 18$ and $2 \times 9 = 18$. Both have an area of 18 square units. The other pairs give different areas.

28 The $\frac{1}{4}$ slices are bigger because the pizza is cut into fewer pieces. More equal parts means smaller pieces: $\frac{1}{4} > \frac{1}{8}$.

29 The letter "L" has one corner where the two lines meet at 90°, forming 1 right angle.

30 An isosceles triangle (2 equal sides) has 1 line of symmetry, running from the top vertex to the middle of the base.

Great job checking your work!

Keep practicing and you'll be a math star!

Find more at
ViewMath.com/FL-Grade3

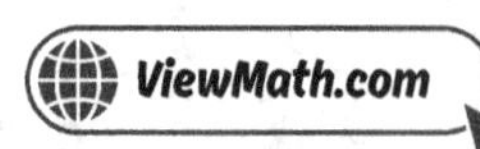